I0059391

The
BIOSPHERE

Rebecca Woodbury, Ph.D., M.Ed.

Gravitas Publications Inc.

The BIOSPHERE

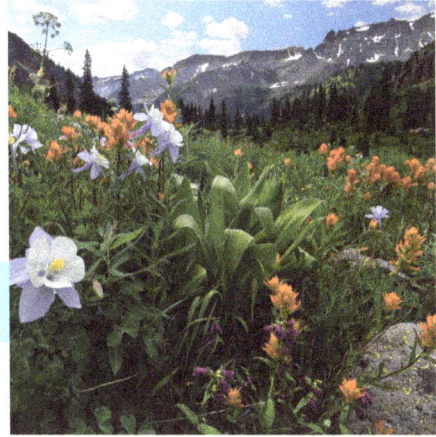

Illustrations: Janet Moneymaker

Copyright © 2025 by Rebecca Woodbury, Ph.D., M.Ed.

All rights reserved. No part of this publication may be reproduced, stored in a retrieval system, or transmitted, in any form or by any means, electronic, mechanical, photocopying, recording, or otherwise, without prior written permission from the publisher. No part of this book may be reproduced in any manner whatsoever without written permission.

The Biosphere
ISBN 978-1-953542-20-5

Published by Gravitas Publications Inc.
Imprint: Real Science-4-Kids
www.gravitaspublications.com
www.realscience4kids.com

RS4K

Photo credits: Cover & Title Pg: julia_arda, AdobeStock; Above: Glenn, AdobeStock; P.7. SunnyS, AdobeStock; P.9. Top, Qrisio, AdobeStock; Bottom, Matthew Field, www.photography.mattfield.com, CC BY SA 3.0; P.13. 1) Thomas from Pixabay; 2) Michael Connor Photo, AdobeStock; 3) vlad61_61, AdobeStock; 4) Glenn, AdobeStock; P.17. Top, ondrejprosicky, AdobeStock; Bottom, Leoniek, AdobeStock

Earth is more than just rocks, water, and air. Earth also has trees, frogs, bugs, rabbits, elephants, and flowers.

The **biosphere** is made up of all the living things on Earth.

The biosphere includes plants, animals, bugs, and all other living things. The biosphere also includes the land, water, and air where living things exist.

Different parts of the biosphere work together to support life.

Soil provides the water and nutrients plants need to live. Animals drink water and eat plants and other animals for food. Birds fly in the air and catch bugs to eat. Animals breathe air and plants use air to help make their own food.

Have you ever noticed that all plants and animals do not live in the same place?

Each type of plant and animal lives in a place that has what it needs to grow and survive.

Plants make their own food from sunlight, air, and nutrients in the soil. They need to live in a place where they can get the right amount of each of these.

Animals need to live in places where they can find the right kind of food. Some animals get their food by eating plants. Some animals eat other animals. Some animals eat both plants and animals.

Some animals eat cheese!

Plants and animals need to live in a place that has the right amount of water and the right kind of weather and temperatures.

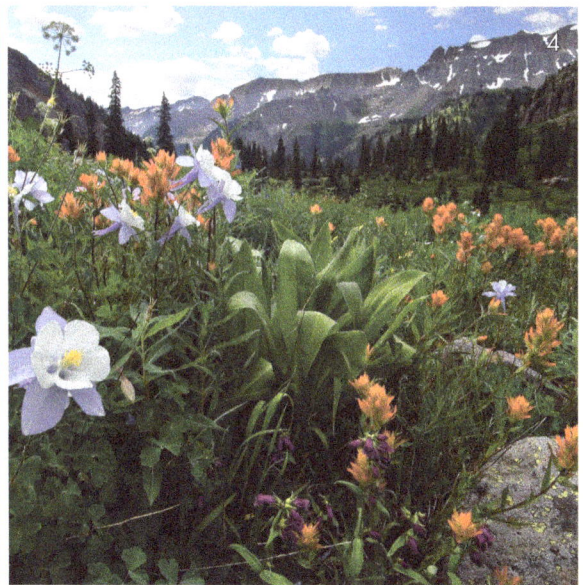

An **environment** is everything that surrounds a living thing in the area where it lives. Water, weather, plants, and animals are all part of an environment.

Would you like to live in a desert environment?
Not really.

A hot, dry desert area is one type of environment. A damp, shady forest is another type of environment. Different plants and animals live in each type of environment.

I like the forest!

A **habitat** is the area within an environment that has all the things a particular plant or animal needs in order to live. Different plants and animals need different habitats.

A squirrel's habitat

We can help keep plants and animals healthy by learning about their environments and habitats and how the different parts of the biosphere work together.

How to say science words

biosphere   (BIY-uh-sfeer)

desert   (DEH-zuhrt)

Earth   (UHRTH)

elephant   (EH-luh-fuhnt)

environment   (in-VIY-ruhn-muhnt)

habitat   (HAA-buh-taat)

nutrient   (NOO-tree-uhnt)

science   (SIY-uhns)

temperature   (TEHM-puhr-choor)

weather   (WEH-thuhr)

www.ingramcontent.com/pod-product-compliance
Lightning Source LLC
Chambersburg PA
CBHW040152200326
41520CB00028B/7581